Drones

Unmanned Vehicles, UFO's and Government Use

2nd Edition

by Brenda Foster

By reading this document, the reader agrees that under no circumstances are we responsible for any losses, direct or indirect, which are incurred as a result of the use of information contained within this document, including, but not limited to, —errors, omissions, or inaccuracies.

Contents

Introduction

Drones have been around since the eighteen hundreds. They are found in the airspace throughout the world and are more popular now than ever before. They were originally used for military action but are now used for commercial and personal use.

Drones are used to watch for potential drug deals, cattle thieves, and even illegal immigrants trying to cross the border.

The Washington Times recently reported experts expect as many as 30,000 drones in the airspace by the year 2020. Not everyone is pleased with that statistic. Many worry that their safety and privacy may be compromised. Some fear that the government will use drones as a spy tool to spy on citizens. Others fear that citizens, now having access to personal drones, will also use them inappropriately.

The advantages to the use of drones in the military are lengthy and covered throughout the book. There are also many commercial advantages to the use drones. Surprisingly, the personal use of drones is quite common and the reasons for owning one are discussed.

Rules and regulations are an ongoing topic throughout both the United States and the rest of the world. The FAA and legislation is working to keep up with the quickly advancing technology.

Is your privacy at risk? Is safety a concern? Just exactly how are drones being used? Get an unbiased glimpse into the world of drones.

History of Drones

Technology found around the world continues to progress. Some of the elements that are created from it change the way we view things, forever. One of those inventions is the drone. Early drones were referred to as unmanned aerial vehicles (UAVs).

A drone is able to be recovered after a mission. Some drones can carry ammo and fire at a given target. Other drones take pictures or surveillance video.

Austria Attack on Venice

The first recorded use of UAVs was in 1849 in Austria. They were used to attack Venice in the form of dropped balloons. Many of them were successful, but the Austrians didn't account for the wind. Some of those balloons of explosives actually traveled back to Austria before they exploded upon impact.

World War I

In 1916, UAVs were used in World War I to carry weapons to various destinations. They were fast and often referred to as aerial torpedoes. There were plenty of designs in the works for them through 1918. However, the United States government didn't want to be too reliant on them during the war.

World War II

In 1935, a prototype was introduced to the US government by Reginald Denney. Approximately 15,000 of these drones were manufactured and used in World War II. By 1938, they were commonly used by both the US Navy and the US Air Force.

Nuclear Testing

In 1946, drones were used for a variety of types of nuclear testing. Explosions were created and different prototypes of drones were sent through the destruction to see how much damage they could withstand. It is believed that the Soviet Union also conducted such testing around the same time period.

Vietnam War

Drones were a huge benefit in the Vietnam War for the US. It is believed that only about 6 of the drones used in this war were able to be shot down by the enemy.

United States Customs and Border Protection

Since 2007, drones have been used by the US Customs and Border Protection. They have been used in an effort to reduce illegal immigrants from crossing the border. It is also in place to assist in reducing the risk of drug trafficking at the border.

United States and Afghanistan

Unmanned drones were used by the US over Afghanistan beginning in 2002. This was during the hunt for Osama bin Laden. The decision to use drones in such a manner was part of the updated security plans after the horrific terrorist attacks on the United States, September 11, 2001.

Emergency Response

In 2005, drones were requested to look for people who may need to be rescued after Hurricane Katrina. However, the FAA didn't have such authorization in place yet. That authorization rule was changed in 2006 and drones can now be used for future emergency situations.

Emergency situations include searching for people trapped or unable to reach a safe location after a natural disaster or after a vehicle accident. Many forest areas and mountain terrains are too dense and dangerous for airplanes. The use of drones can rescue and save lives quickly.

There has been the suggestion to use drones to help look for people in avalanches or lost in hiking expeditions. A great deal of time and planning is required to implement a search and rescue team to get to those often remote locations. Drones can get there quickly and easily to help rescue and potentially save lives.

Advantages

Drones continue to be used because of the many varied advantages they offer. Noise and privacy issues are common complaints. However, what they offer is well beyond such complaints.

Price

Drones can be expensive. However, the cost is far less than the equipment and manpower that the drone is able to replace. Drones are also faster and less expensive to build than other aircraft. If they are they be damaged or destroyed, the loss will be less. Drones can be made in smaller sizes, reducing fuel usage. Many are now solar powered.

Human Safety

Manned aircraft can be compromised, putting the humans on board at great risk. Aircraft that is compromised can also be used to cause harm to other people or to structures. Drones are very hard to compromise. The worst case scenario is that they get shot down and destroyed. However, the enemy isn't able to capture them and use them for their own personal gain as they can with an aircraft.

Drones are safer than sending men or women into various dangerous situations. The military has utilized this valuable aspect of the drone in order to protect serviceman and women. Drones can also fly into tighter areas, reducing the risk of an accident or a crash of a manned aircraft.

Drones are unmanned, and when they are destroyed or shot down, there are no lives lost. The number of drones that were shot down in World War I and World War II was very low. However, the number of them shot at and destroyed over Afghanistan was extremely high.

Manpower

Drones can replace the manpower of a large number of men and women. The savings can add up very quickly. Not only in terms of wages for those employees but also for transporting people, food, and lodging.

Reduces Errors

Weapons and the use of drones can also be safer. High stress situation occur during times of war. Using drones to carrying weapons reduces the risk of human error under stress. This includes releasing weapons too soon, without authorization, or over the wrong targeted location.

Testing

Being able to test drones in different situations allows for better construction of other aircraft. It helps to identify safety concerns and problems that otherwise may not have been identified until a serious disaster has taken place.

Secrecy

Due to the small size of some drones, they are often difficult to detect. This is important when there are military operations in place, possible drug trafficking issues to verify and stop, or possible crimes taking place.

PTSD

Many men and women in the military suffer from PTSD (Post Traumatic Stress Disorder). Combat is the number one reason for PTSD. Many experts believe that by sending in drones to take part in what they call the "drone war" zones they can reduce the effects of PTSD on military personnel.

Hours of Operation

There are only so many hours of flight time that humans can cover before they must take a break. Their bodies and their minds simply can't continue to be alert and function without sleep. Drones can stay in the air and around their target area without any breaks. Since they are unmanned, they can stay in place for days, weeks, and even months or years if they have to.

Improved Structure Inspections

Routine inspections should take place for bridges and buildings and can often be dangerous to complete. Time and safety cost states a great deal of money for routine inspection and repair work. The use of drones to review structures from all angles and to take photos is a great alternative. The work can be done in less time, with a great deal of accuracy, and without putting workers in danger.

Commercial Use

Confusion about the commercial use of drones continues. They are rules and regulations but they are often hard to clarify in some areas, allowing them to be used commercially.

Drones can legally be used as a hobby. The owner is required to keep the drone within their sight at all times. Why would someone use a drone? Taking photographs to get a terrific aerial shot would be one purpose for a drone.

Commercial businesses get around regulations by setting up a non-profit business or stating that they are a hobbyist. Both of these classifications allow them to operate drones without being considered a violation of commercial use.

They are able to collect payments, too. However, money has to be collected in a manner classified as donations. While the FAA is aware of this, they haven't made it a priority. They haven't investing a great deal of time or effort to stop the practice.

The FAA regulations state that drones can't be used for commercial gain. However, the US Congress has required the

FAA to change this. Beginning in 2015, the FAA will have to allow drones into commercial airspace.

It was announced 12/30/13 by the FAA that six states will be able to develop testing sites for drones. Those states are:

- Alaska

- Nevada

- New York

- North Dakota

- Texas

- Virginia

The reasons that these locations were selected are due to the variations they offer. The FFA considered the following different types of locations:

- Airspace

- Climate

- Terrain

According to the FFA, they believe this is a huge step towards gathering sufficient and reliable data. They feel this information will assist them with creating rules and regulations needed for commercial drone use in 2015.

The testing sites are anticipated to generate an undisclosed number of new jobs in those states. Not everyone is happy about this, though. Many want drones to be approved for commercial use without all the red tape of rules and regulations.

A survey in 2013 showed that 42 states agreed to bills being implemented in their states that restrict the use of drones. The restrictions are mainly regarding privacy concerns.

There are eight states that have already imposed such bills. Those states are preparing for the FAA changes in 2015. It is expected that many more states will also be getting these bills acted upon and implemented before the commercial use of drones is accepted by the FAA.

Too Many Drones

One of the biggest worries that the FAA has about commercial use of drones is that there will be too many in the airspace. There has been talk about using drones for fast deliveries by some of the larger retailers including:

- Amazon

- Sears

- Wal-Mart

This could literally result in hundreds of drones in the airspace, as the orders are in route to their destinations. Drone numbers could dramatically increase around the holiday season, increasing concerns.

Obviously, FAA regulations will require caution and diligence. Putting rules and regulations into place will help ensure that there isn't chaos from the commercial use of drones.

The most common commercial ventures trying to get approved for drone use include:

- Business Logistics

- Insurance and Property Valuation

- Journalists

- Motion Pictures

- Oil and Gas Companies (Pipelines)

- Photography

- Real Estate

The experts believe that there will be growth in 2015 in the area of commercial use of drones. However, they believe lots of commercial entities are going to be disappointed and frustrated by the lack of freedom they have.

Manufacturers

From a manufacture's point of view, the ability of commercial use is very exciting. Manufactures have already started with the development and promoting of consumer drones. They are hopeful they can generate a great deal of profit by selling drones to the commercial sector.

Some of the top manufacturers for commercial drones include:

- **DJI Phantom** – This is considered to be the first mass market drone. There are four configurations that consumers can pick from. They are all designed to be very simple to operate and range in terms of the weight they can handle. The Quadrocopter is one that is getting a great deal of attention. They also offer the 2 Vision Quad.

- **Parrot** – This is a very small drone with a configuration that looks very similar to a butterfly. It is also very cost efficient, which is perfect for a business that doesn't want a huge investment in this type of marketing.

FAA Fight

The FAA isn't going to give up without a fight. In March of 2013, they appealed a Federal court ruling regarding the use of drones for commercial needs. The FAA is not pleased that the charges against Matt Murphy in January of 2013 were dropped.

The charges against Murphy include a $10,000 fine by the FAA for recklessly flying a drone at the University of Virginia for a photography shoot. The charges were dropped due to the fact that there were no enforceable FAA rules and regulations at the time. With this in place, the precedence in the court of law has been set. The FAA has to come up with additional rules and regulations so that what they deem as violations can be prosecuted.

The outcome of the FAA appeal regarding the use of drones for commercial use as being legal is one that will be watched by many. There may be some compromises made in the end for all parties involved.

Recent Military Use

Al Qaeda

Drones were used in 2002 against Al Qaeda in Yemen. In 2008, they were significantly used by the Bush Administration against Pakistan. The Obama Administration has continued to use drones against Al Qaeda.

The CIA is in charge of a great deal of the drone use against Al Qaeda. However, elite groups use them for military strategies and maneuvers. One elite group is the Joint Special Operations Command (JSOC). They have mainly relied on drone use in Yemen and Somalia.

The CIA, JSOC, and other military units compile what they refer to as the "kill list". Meetings are held and the list has to be approved by the Pentagon. Once that is completed, the list has to be approved by members of the White House cabinet and the President.

The National Counterterrorism Center often sends recommendations for who they think should be added to the list. While this "kill list" is quite controversial, many experts believe it is a necessity. It is a way to keep people safe day after day. No one wants to allow a terrorist attack like we experienced on 09/11 to occur again.

Israel Concerns

Israel has concerns over Gaza and Lebanon being able to send exploding drones. They already have arsenals of rockets that they can use against Tel Aviv. The Chief of Israel doesn't like the idea of being in a position where they have to cope with numerous AUVs. They fear attack from both the South and the North.

Drones have been reported on two occasions to date in the airspace of Israel. One of the occurrences was in 2012 and the other was in 2013. It is believed that they were carrying cameras, not explosives, in an attempt to visualize Israeli defenses. Both drones were shot down by Israeli jets.

Navy

The use of drones can help to improve security as they can be launched in areas quite a distance from the actual naval ships.

Coast Guard

The US Coast Guard believes that the use of drones can help increase prosecutions by as much as 95%. Drugs are being pushed through the waters and the Coast Guard needs the drones to assist them with making drug busts and breaking down that illegal entity.

However, there are strict restrictions and the Coast Guard shares their drones with the Navy. Some of the projects in place will allow them to use the drones up to 70% more than they are right now.

Drones and the War on Terrorists

Drones have been a favorite weapon of the Obama administration as a safe way to fight the war on terrorists. They have been significantly involved in approved airstrikes both in Iraq and Afghanistan. An undisclosed analyst for the military states that up to 95% of direct target killings in these countries since 2011 have been conducted with the use of drones.

Drone Data

Some people worry that the use of drones in the military are going to replace the number of people that they employee. That isn't really the case, but it can change the types of situations that the humans in the military operations are found in. For example, there are anywhere from 65,000 to 70,000 military personnel that work on compiling, interpreting, and using the data from drones.

According to a review of military operations, it is believed up to 100,000 personnel need to be dedicated to this particular type of work. As the number of drones in use increases, so will the number of people that they allocate to such tasks.

Conflict of Drone War

There are plenty of countries out there that don't like the concept of what is dubbed a "drone war". Some countries have been protesting the use of drones. The Pakistan militia has a campaign in place that says the US is spying on them. They also report that the US is killing innocent civilians.

A report compiled from the United Nations (UN) in 2010 also raised some concerns. Many allies of the US are hesitant to speak at all about the use of drones. They don't want to get into a conflict about it. The conclusion is that there is a lack of good information about drone use in the military. It has made many people, and even leaders of various countries uneasy.

Other Uses

There are plenty of other possible uses for drones that the average person may not have thought about. As you read through these, you may identify a couple of them that you would be glad to consider implementing for your own personal use.

Cinematography

Both amateurs and professionals enjoy the idea of being able to use drones for cinematography. This is the art of taking videos or action shots for movies or videos. Drones would be able to get up close to the action and really capture the scene.

Such images could completely change what is used in the final production. It could add significant detail to what the viewer will be able to identify with in any given scene. The use of drones in films that capture animals in nature would also be explosive. It would allow them to be filmed in their natural habitat without any risk to humans going in and setting up cameras.

Drones are replacing helicopters for these types of shots. Drones are quieter than helicopter, don't disrupt the landscape, and can take picture from great distances.

Action Sports

Many people find action sports to be very entertaining. Instant replays is wonderful, but doesn't always capture the details that can become controversial. A drone could be able to get up close and personal without being an invasion of space for the athletes.

Some action sports have a referee with the athletes. They have to be close to see the action, but they also have to protect themselves at all times. They have to be ready to move quickly out of the way with the movements of the athletes. The use of drones could allow those referees to be a safe distance away from the action.

Real Estate

Even though commercial use of drones isn't legal, it still happens. Drones are popular in real estate. Being able to take photos of properties from all angles is important. Capturing the home and the surrounding property from the air is a great way to capture the attention of potential buyers.

Photography

Both amateur and professional photographers use drones to capture unique photos. Drones help take photos of those hard to reach places. Or can give interesting points of view.

Agriculture

Farmers and ranchers are using drones to help them with their business. Drones are being used to help identify problems with crop growth or irrigation. Or to monitor large areas of land.

Ranchers can use drones to count cattle and look for those that may have been separated from the herd. They can also use drones to help them identify problems such as cattle theft, breaks in the fence, trespassing, etc.

Both farmers and ranchers have to invest time and money in the upkeep of their facilities. The use of drones can be a reliable tool for surveillance and monitoring of their land. There are over 10,000 drones in use in Japan for agricultural purposes.

Law Enforcement

Many areas of law enforcement can benefit from the use of drones. For example, looking for suspects that are on the run, whether on foot or in a vehicle. Drones can monitor known shipment areas for drugs. They can monitor traffic on interstates and help keep traffic flowing safely.

Weather Observations

Drones can be used for weather observations as they can get photos from different angles very close to the storms. Drones can get close to the eye of a hurricane or very close to a tornado, for instance. These drones are often destroyed, but they can provide a tremendous amount of data and information for us to study and learn from.

3-D Mapping

With the use of drones, 3-D mapping is becoming extremely popular. This process allows the landscape to be carefully surveyed. Thousands of images can be taken in a small amount of time. The drones can produce better results than satellites for creating these 3-D mapping elements.

They can be guided by GPS so they don't have to be manually guided by a human. One of the main sources for this type of 3-D mapping is Pix4D out of Switzerland. The images they had captured help with the Haiti relief efforts following Hurricane Sandy.

Wildlife Protection and Documentation

Many forms of wildlife are in danger of becoming extinct. Environmental issues, poaching, and other factors make it difficult. Drones are a non-invasive way to protect wildlife and to document where their habitat continues to be found. It can also help estimate numbers that still remain in the wild.

Drones can map roads that allow them to reach wildlife in order to offer help, add monitoring devices, and to install cameras around a given habitat area. Drones can be used to help identify poaching activities and even capture those responsible due to the images that can be recorded.

One of the huge efforts in place right now is for conservation of Orangutans in Malaysia and Indonesia. The use of drones has proven to be far more effective than efforts on ground due to the thick forest of their habitat.

Privacy & Safety Concerns

While there are plenty of benefits and uses for drones, not everyone is happy with them. Many people have concerns about their privacy being violated and question the safety of drones.

Privacy

Many feel drones are going to invade their privacy and report their personal activities. They fear they are being watched by the government. They feel it is similar to having a surveillance camera on every corner. Most value their privacy and feel that drones are an invasion of privacy.

Some people think that drones just look unfriendly. They see them as a threat. One of the elements that many manufactures are working on is their visual attractiveness. They are hopeful that those opposed to them won't mind so much if they are brighter in color or more visually appealing. However, there are those that argue it is the privacy issue and not appearance that they are upset about.

Criminal Use

Criminals always seem to be able to find a way to use devices to their benefit. There is the chance that criminal would use drones in order to help case homes or businesses that they planned to rob.

However, the flip side of it is that with drones hovering, crime rates may actually go down. Criminals look for a window of opportunity most of the time where there is little risk of getting caught. They look for easy targets, such as people walking alone, and homes that look dark and unoccupied.

Drones in the air could increase the chance of a robber getting caught in the act or caught on video. Therefore, drones could help decrease certain crimes in some areas.

Personal Privacy

Some worry that small drones can interfere with their personal privacy. This includes:

- Taking a shower

- Using the bathroom

- Getting dress/undressed

- Intimacy

- Conversations

- Work activities

Some have a fear that drones could literally be everywhere, monitoring everything they do. They simply want their privacy in their daily lives and fear that giving up any rights will lead to giving up all rights.

Safety

Drones are made of carbon fiber and are very lightweight. Severe winds could cause them to crash into a person, a vehicle, or even onto a busy highway.

Animals may be scared or harmed by drones. The sound causes some animals to run or react in odd ways.

Noise

Depending on where you reside, you may be used to high levels of noise. However, others aren't and they view it as a type of noise pollution. They also feel that their privacy has been invaded due to the noise. While a single drone in an area isn't going to be excessively noisy, people are concerned about the increased number of them over time.

People have complained about drone noise to their local officials, to state representatives, and even to the Federal government.

Crowded Airspace

According to the FAA, there are already concerns with drones in the US due to crowded airspace. There are fears that the addition of drones in the US in large numbers could further cause problems. Crowded airspace could result in flight delays, problems with detection of what is in the airspace, and more.

Privacy Legislation

Many states are looking into privacy legislation relating to drones. They are doing this now in the event that the commercial use of drones is able to be opened up in the near future. Such privacy issues involve limiting how drones can be used. If they are used for criminal acts or invasion of privacy, very stiff penalties should be in place.

Many states believe that by getting legislation in place, they can eliminate the fears that have many concerned that their privacy is going to be invaded due to the use of drones on a wider scale.

FAA Concerns

Privacy is also one of the concerns of the FAA. They have authorized 78 certificates for commercial use. They have already had to increase their staff to take care of those licensing issues and to address safety concerns from citizens.

The FAA is aware that many law enforcement entities in the US have purchased drone equipment. This includes the states of:

- Alabama

- Florida

- Texas

- Washington

- Virginia

These states are also addressing security issues as well as those complaints that the FAA has seen. They want their citizens to know they will be safe with the use of the drones in place. In fact, they enjoy the fact that they allow data to be collected in real time. Drones can be reviewed in terms of camera footage to gather information for crimes.

Such information can be instrumental for law enforcement to cut down on the time needed to conduct an investigation. The photos can also give them documentation and evidence they otherwise would not have been able to have in order to successfully prosecute someone for their role in a crime.

The FAA says they have been cooperating with efforts by US states that want to use drones for law enforcement. They have been doing so since 2010. However, they want security measures in place regarding who is going to be responsible for observing the drones and how they are used. The risk of that authority falling into the wrong hands is what the FAA wants to work to prevent.

The Department of Homeland Security is now in that mix, too. In fact, they are currently looking into offering grant money for those states with law enforcement that would like to buy drones but can't afford them. Their mission is to keep citizens safe. This includes locally and on a national level from criminals and from terrorists.

They hope that citizens would be willing to give up a small amount of personal privacy in order to live in a location where there is less crime, less drug activity, and less risk to everyone.

The Future of Drones

Drones seem to be here to stay, not just a passing trend. As mentioned, the regulations are changing for them to be used commercially. They are used widely by the military and they are also used by private citizens. The demand for them will likely increase in the future.

Fail Safe Failure

Drones continue to be very advanced. Personal drones have a fail safe button. When the button is pushed the drone is supposed to return to the takeoff spot. But what if it doesn't? Who has the ownership rights to drones that end up in private locations?

Recently, a man in Denver realized his drone had drifted away. The fail safe button didn't work; the drone never returned. The value of the drone was $2,000 plus $400 for the camera installed on it. All he was left with was the transmitter. He feels like the company owes him an explanation.

He isn't the only one with such a complaint. There are numerous forums and reviews online about this very issue.

With this negative publicity out there, manufacturers of drones have no choice but to improve their product design. They don't want sales of these devices to slow down due to people worrying about their investment getting away from them.

Commercial Use

We touched on this in a previous chapter, but it is worth addressing here, too. The future of drones is likely going to really grow in the commercial sector in the coming years. This is due to the fact that the FFA regulations that ban them from such use right now will be either modified or lifted.

Price

It is anticipated that the price tag associated with drones will decrease around 2015. If that is true, then there will likely be more of them in use. For some consumers, it is cost that prevents them from making the investment.

Protecting the Boarders

Illegal immigration from other continues into the US continues to be a serious problem. Drones are used along with manpower to protect the boarders of the US. Some pilot programs are in place, a few of the drones have actually crashed. Improvements are needed if this resource is going to be a reliable method of keeping illegals from crossing the border.

Projections

By 2017, it is estimated that there may be as many as 20,000 small drones.

There are about 56 different government agencies that use drones. There are 63 active drone sites where they can be used. This doesn't include the 6 testing locations that were mentioned in a previous chapter. There are plenty of experts that believe the 21st Century is going to be the era of drones rather than manned aircraft.

Little Known Facts About Drones

Drones have actually started to become an integral part of our modern world. As things move towards increased and unparalleled mechanization and automation, it is only logical that drones will begin to possess an increased presence. What this means is that, as of late, there has been an increase in drone use that borders on alarming when one considers the fact that they are being used for purposes both benign and not so benign.

However, what do we really know about drones? There are so many things that seem important but are actually never really thought about. Where did drones come from? When were they first put to use and for what purpose? The answers to such questions can potentially reveal a lot about these interesting, useful and occasionally sinister machines.

A lot of people would say that what drones can potentially do now, their uses that border on miraculous, should be enough to make us ignore the past and focus on their future. However, it is only in the acceptance of these little known facts about drones that we can begin to understand the parameters of these devices that can give us so much and, at the same time, take so much away from us. What follows are the various little known facts about drones.

The First Military Use of Drones

Drone technology has existed in speculative scientific conjecture since perhaps the 1970s, enjoying a minor presence in science fiction during the sci-fi craze of that era. However, the use of drones was considered relatively tame until the 1990s, when the world found itself faced with a new public enemy that had to be stopped at all costs.

This enemy was Osama Bin Laden, a terrorist and extremist who had begun to make a dangerous mark on the world. Considering that the terrorist group he started, Al Qaeda, would go on to commit one of the biggest atrocities in the history of the world, the bombing of the World Trade Center on the 11th of September in the year 2001, they were, perhaps, the prime candidate on which new and improved surveillance technology could be used, and it is to the United States Military's credit that they had the foresight to use this technology.

In the year 1998, following an attack on an Al Qaeda camp that resulted in three hundred civilian deaths, the US

Military and the President of the United States at the time, Bill Clinton, decided that a more precise method of surveillance and remote operated offense was required.

Hence, the Predator drone was brought into being, and it was first used in a cover operation that attempted to apprehend Osama Bin Laden. It was wildly successful, and offshoots of the Predator drone are still used today to target terrorists in Northern Pakistan and Afghanistan. These drones allow the targeting of terrorists without putting innocent soldiers at risk.

Conversely, drones also pose a risk to a country's sovereignty. A country like Pakistan, for example, which suffers from a severe terrorist threat, is not respected. Its borders, its sovereignty, its ability to govern and police its own borders are disregarded when a foreign military bombs its soil. The use of drones makes such a gross violation of sovereignty possible because drones are usually compact and not very easy to detect, which allows foreign militaries to use them for surveillance and covert operations all the time. Such is the cost of the creation of such a revolutionary technology, but it is a cost that we must bear in mind moving forward.

Drones Are Not As Effective As You Think

People often think of drones as these unstoppable machines akin to what we have begun to see in speculative, apocalyptic science fiction such as the movie "The Terminator", in which sentient machines wipe out humanity. The idea of a drone being an impossible to stop machine is actually quite laughable, when one considers how likely it is for a drone to crash or stop responding completely on a covert operation.

The fact of the matter is, drones are sent out on missions and are remotely controlled, but they are not human. They have no human responses or reflexes, and so any communications that they send back are bound to be flawed. The sensors that are present on a drone can send back objective data, but there is no chance of speculation. It can be argued that a drone is controlled by a soldier and so the soldier's ability to reason is present within the drone, but the soldier will be acting on incomplete information at all times. There will be no peripheral vision, no tingle on the back of the neck when they feel something is wrong, and no bad feelings that can prevent them from walking into a terrible situation.

Hence, drones often crash due to bad weather, enemy attacks or just simple malfunctions. In fact, the United States has crashed around eighty drones so far, a shocking number when you consider that the total number of drones that the United States military has flown so far is still in the early hundreds. This means that one out of every five drones is going to crash, this much is a certainty, which casts some serious doubts onto the efficacy of drones in general.

Will Be Part of the Police State

Of late, there has been increasing concerns regarding the militarization of the police of the United States. Police departments now possess increasingly dangerous and heavy duty weaponry, the likes of which may well be used to fight terrorists rather than to apprehend criminals whether petty or major.

Additionally, as of late the police departments of several cities have been accused of any combination of racial profiling, unnecessary use of force, misuse of authority and lack of the ability to judge whether deadly force is necessary in a situation or not. To top all of this off, police departments are being accused of irrationally and unconditionally defending their police officers, men and women that are often accused of murdering innocent people in cold blood as has been seen so many times over the past few years. What this creates is an entity that possesses significant resources, heavy weaponry, extraordinary manpower and zero accountability, an entity that, in any other situation, would have been eradicated or at the very least completely revamped from the bottom up by the government.

As a result of these alarming trends in various the police departments of the United States, a lot of people have begun to posit that the United States of America is actually a police state or is well on its way to becoming one. The misuse of power in present circles has lead many scholars to speculate that, even if the police state is not currently being enforced, it is well on its way based on the trends that are being seen in policing across all major cities in the United States of America.

Whether the police state is here yet or not, one fact can be accepted without any question: drones will be a part of the police state. The NSA was recently revealed by the whistleblower Edward Snowden to have been using its resources to spy on American citizens through their smartphones, computers and on social media. All military entities within America are placing great emphasis on the gathering and procurement of domestic intelligence, ostensibly to prevent any terrorist threats to American citizens that are homegrown by foreign agencies through the use of sleeper units, undercover terrorists and foreign agents that are waiting for the right moment to strike.

Drones are going to be a huge part of gathering this intelligence on American citizens. Already, the dystopian doomsayers are proselytizing about how the government is going to police us using drones, monitor our every action and probably prevent us from committing crimes.

This is backed up by the fact that hundreds of certificates for the use of over eighty various types of drones were acquired by various military and intelligence agencies in America, with about a third of these being acquired by the Central Intelligence Agency and a significant chunk going to Homeland Security.

What these drones will be used for, no one can say. But the fact remains that America is becoming increasingly policed, its citizens are gradually losing their civil liberties without even knowing it, and the fact that military and intelligence agencies are attempting to operate these drones on American soil is probably a good indicator of how much surveillance and policing will actually be conducted on American citizen in the not so distant future.

Drones Aren't One Trick Ponies Anymore

Gone are the days when the Predator drone was the only drone that people knew about. Gone are the days when the only thing drones were used for were covert missile attacks on terrorist bases. Those were, admittedly, simpler times, but the need for more advanced drone technology has spurred the creation of several different types of drones, each with a distinct use in the world of military operations, each just as effective as the last in its specific field.

Soldiers have a difficult job, to be sure. They risk their lives doing what they do, and so the armies that they are a part of will inevitably try to stop them from having to risk their lives. The creation of machines that can do their job for them was a huge boon to these soldiers, and as a result a lot more of said machines have been developed.

Drones are being used for simple surveillance. These drones are usually delicate and extremely stealthy, with their entire design being based on the single idea that they must be difficult to spot and impossible to hear. These drones also possess specifications that make them extremely difficult, if not utterly impossible, to spot or detect via radar technology.

There are a huge array of other drones as well, each of which are just as important as the last. The polar opposite of stealth drones is, perhaps, a much bulkier type of drone that is used to break down doors. This drone is probably a lot easier to see, but that's because it needs to be the size that it is in order to facilitate the job that it is required to do. These drones are also often used to provide injured soldiers with cover. Their bulkiness and overall solidity and strength make them excellent barriers and shields, things that are invaluable in combat situations.

Although these drones can be used for other things, they are intended for a single purpose. Using them for any other purpose is innovative and requires thinking on your feet, but will invariably result in the drones getting damaged in some ways. However, a lot of drones that are being created these days are created with the express intention of being multifunctional in the first place. In a combat scenario, having a lot of single purpose tools is not as effective as having a single tool that can accomplish a variety of tasks. An example of such a tool is a Swiss Army Knife that has multiple tools any of which can prove useful in a combat scenario.

As a result of these advancements in drone technology, the use of drones is becoming increasingly common in military situations. It is not all that farfetched to assume that wars in the not too distant future might just be fought between drones, with human involvement being limited to the controlling and aiming of said drones. Although this would certainly greatly limit the amount of casualties that armies would suffer, it would also detach soldiers from the collateral damage they are causing, something that doesn't bode very well for the people that would be suffering due to this collateral damage.

Civilian Use of Drones is Also Increasing Drastically

It is widely believed that the use of drones is restricted to the military. After all, when one thinks of drones one imagines a cold machine that is on its way to a terrorist base to launch a missile and destroy said terrorists. This is also the most common depiction of drone technology in popular culture, particularly in TV shows and movies depicting military operations in countries prone to terrorist activity. When looking at all of this ostensible proof that drones are used primarily by the military, it is only natural for one to believe that this is how things are in the real world.

However, the idea that drone technology is exclusively or at least mostly used by the military is a huge misconception. Drone use in the public sector has increased drastically in recent times, with many companies realizing the potential of these machines in helping make day to day tasks easier than they would be otherwise. Although it is true that drones were initially brought into regular use by the military, the fact remains that the people that have nothing to do with the military are far, far larger in number than the people that are in it or actively involved in it.

One of the biggest examples of the appropriation of drone technology by a civilian entity is the initiative started by the online retailer Amazon in which it would provide goods to its customers within an hour through the use of drone technology. Although Amazon's drone technology has not yet been perfected, and there has been a slight problem acquiring permission from US military to use certain airspaces over particular cities, the initiative proves that there is significant potential in the use of drone technology among civilians as well.

Drones are also frequently being used by journalists and artists to capture photographs. In a less positive take on the aforementioned use, paparazzi are starting to use drone technology to capture photographs of celebrities, enabling them to take photos that would have impossible to take before. This tells us an important fact on how drones can be used to invade privacy. However, on a more positive note, drone technology is also frequently being used by environmental activists to prevent illegal whaling and to monitor endangered species without being invasive or intrusive, something that puts a positive spin on the stealthiness of drone technology.

Drone technology has also begun to be used in an increased capacity in the private sector as well. Civilians have begun using drones for their own, private use as well for a variety of reasons. From personal security to more unseemly uses such as spying on their neighbors, people have begun to pay exorbitant amounts of money in order to acquire drones for themselves. What all of this means is that drones are becoming extremely important parts of our lives, and will very soon become exponentially more common than they are at present.

Military Uses for Drones Go Beyond Just Bombing

Once again, the general concept of drones that is pervasive in society today must be discussed. Even after reading the above statement, one can still think that, despite the fact that drones are just as common among civilians, if not more so, as in the military, the way drones are used in the military is accurately represented in the media. Drones are used to bomb the enemy, they are used to remotely launch deadly weapons onto enemy bases so that soldiers don't have to attack the base themselves.

It is true that the first ever use for a military drone was the bombing of a potential terrorist base. This has probably lead a lot of people to believe that this is still the primary use for drones. Additionally, the only times that drones are actually mentioned in the headlines is when they have been used in decapitation strikes, that is when drones are used to bomb enemy camps. This is probably due to the fact that such strikes are the only times when the use of drones actually seems exciting and, therefore, newsworthy. The other times that drones are used are not as exciting and are therefore not commonly reported on.

As a result, the only time the common public ever hears about drones is when they are used to bomb enemy bases. It is only natural and logical, then, that they will associate drone technology, at least as far as the military is concerned, with offensive strikes conducted in a covert manner.

However, as you have probably already guessed after reading the aforementioned section on the vast variety of drones that are now being created, drone technology is being used for a wide variety of other purposes apart from bombing the enemy. In fact covert offensive strikes is actually the least of all of the various uses that the military has for drone technology, which may come as a shock to all of the people that believed that drones are primarily offensive technology.

The major use that the world's militaries have for drones is not offensive or reactionary. Rather, it is more on the defensive and precautionary side of things. Drones are most commonly used for the purpose of reconnaissance, which essentially means the gathering of intelligence. After all, a stealthy machine capable of travelling great distances without getting tired, a machine that allows covert missions to be undertaken without soldiers having to risk their lives, will invariably be used to gather information from places that are not easily accessible.

Drones can be used to take pictures of enemy bases for the purposes of ascertaining what move the enemy might make in the future. They can be used to sense particles in the air in order to determine whether the enemy is making nuclear or biological weapons. All in all, drones are the perfect reconnaissance tool, and are thus used more often for this purpose than for offensive purposes.

The Use of Drones Does Not Necessarily Negate the Necessity of Soldiers

Yet another commonly believed aspect of drone technology is that it saves lives. Drones, after all, can be remotely monitored and activated, they can be flown from a safe location enabling soldiers to sit in the comfort of their bases and not have to risk their lives in service of their country. The logic behind this assumption is sound. After all, isn't that what drones are for? In all representations of drones in pop culture, the fact that they prevent soldier casualties is paramount to their enormous presence in military operations.

However, this assumption, like so many other assumptions about drones, is completely false. Piloting a drone does not require just a single pilot. There is, obviously, the fact that drone pilots work in shifts. Still, none of these pilots are ever in the field. The notion that more drones means less boots on the ground is still valid. However, what a lot of people don't know is that, apart from the people that actually pilot it, drones needs a small army of people that are actually in the field with it in order to ensure that it is doing its job properly.

Over a hundred and fifty people are required to keep a regular predator drone in the air, and the number approaches the two hundred mark when the Predator's larger cousin, the Reaper, comes into play.. These people include the pilots, a maintenance crew, people that monitor the sensors and their readings as well as soldiers that analyze the intelligence that drones provide. When drones crash, whether due to malfunction, bad weather or offensive maneuvers made by the enemy, the maintenance team often have to go into the field to retrieve it or repair it, and a small core of soldiers has to go with them in order to protect them.

Drones actually end up making the army put more boots on the ground rather than reduce the amount of soldiers that are actually in the field. More often than not, drones offer support to actual soldiers rather than do the entire mission by themselves. All of this amounts to one simple fact: drones do not save lives in the slightest. They are military technology, and have greatly advanced the art of warfare but it cannot be said that they prevent soldier deaths because so many soldiers end up dying trying to retrieve them.

The real reason drones are so popular among the militaries of the world is that they are cheaper than actual aircraft that need pilots in the cockpit. Drones cost about a third of what fighter planes cost, and an added advantage is that if they crash and burn, their pilots do not crash and burn with them. Additionally, they are easier to retrieve and repair, meaning that if a drone goes down the military will often not have to build a whole new one in order to compensate for the loss that has been incurred.

Military Use of Drones is Almost Completely Unregulated

Since drone warfare became a common aspect of the global war on terror spearheaded by the American military, international media sources have speculated that around two thousand insurgents and terrorists have been killed solely through drone strikes. Civilian casualties and collateral damage, on the other hand, is completely unknown. Estimates range from as low as a hundred to as high as a hundred thousand. The fact of the matter is, nobody really knows how many people are killed in drone strikes each year due to a wide disparity in statistics.

The vast majority of drone strikes conducted by the American military have been conducted in Northern Pakistan, an area that is notorious for being the home of the terrorist group the Taliban. However, the Taliban have conducted absolutely no attacks on American soil, and their attacks on American assets have been little more than guerilla warfare. They are widely seen as an unruly mob that has no real agenda. Indeed, the vast majority of Taliban attacks have been conducted within Pakistan against Pakistanis, something that is very unlikely to cause the American military to retaliate in such a way.

Taliban presence has also been seen in Afghanistan, the country in which the Taliban were first established before escaping to Pakistan following severe military incursions by America within that nation. Yet drones almost never strike Afghanistan, despite the fact that so many Afghani Taliban are causing havoc in that country as well.

Pakistani statistics estimate that American drone strikes incur major collateral damage and kill tens of thousands of civilians every year and are not an effective means of combating terrorism within that country. American military, on the other hand, maintains that civilian losses incurred by their drone strikes are minimal, and that they are killing actual terrorists with every strike.

There is no consensus about the use of drones within this country, and the far scarier fact is that these drone strikes are being regulated by absolutely nobody. The US military does not release any hard facts about who they are targeting. They are tight lipped about why they are targeting these terrorists as well, an important question to ask considering the fact that these terrorists pose little, if any, threat to American soil or American assets. Their reasons for using drone strikes are their own, and a military agency operating with so much autonomy is something to be feared.

This is not to say that the American military has any unseemly agenda. Their reasons are probably not all that dire, they probably simply just believe that the Taliban are a threat. The reason these things are important because they indicate that the people that are using drones to attack what they refer to as enemy bases have nobody they have to answer to. The fact that they can attack with impunity, and incur collateral damage, no matter how small, must be addressed in order to justify the use of drones in this manner.

Drones Are Not Exclusive to the United States Military

Another generally agreed upon misconception regarding drone warfare is that it is mostly conducted by the American Military. Like all of the other misconceptions regarding drones, the general public's acceptance of this misconception is due to the fact that, in depictions of drone warfare in the media, along with the vast majority of pop culture representations of the use of drones in military operations has shown the US military using drones to bomb their enemies. This has lead to the widespread belief that the militaries of other countries do not possess drones, are not involved with drone warfare or that the drones that they do possess are in some way inferior to the drones that are in the possession of the United States military.

It is true that the United States is the world leader as far as drones are concerned. A majority, albeit a small one, of advances in drone technology have occurred in the US. Three quarters of the procurement of drone research and development has been done by the US, which puts it ahead of the rest of the world as far as drones are concerned.

However, it must also be noted that the proportion of the total budget that the United States allocates towards its military programs is also far greater than the countries in the rest of the world. Taking this into account, the military programs of other countries are making huge strides in incorporating drone technology into their armies.

There are hundreds of drone programs around the world, based in around fifty five countries based on the latest estimates. China has been escalating its drone program drastically, improving it to the point where it is slowly starting to rival the drone program of the United States. In the first decade of the 21st century, China started to secure its place as America's major economic competitor. In the second decade, it has started to become America's only competitor in the military aspect as well, and the improvements it has made to its drone programs reflects this fact.

Another country that America often eyes with unease is Iran. Although Iran's military is not quite as powerful as the military of the United States, its position within the Middle East as well as its nuclear program has made it a cause for concern for the United States, particularly considering the interest that its military takes in that particular region. Iran appears to be creating another cause for concern for the US: its drone program.

Iran's drone program is very advanced, approaching a level that would allow it to conduct regular drone strikes in regions surrounding it. Considering the fact that there are a huge number of American military bases in the region surround Iran, this may be an important factor in the global race for dominance via drones, particularly when one considers the fact that Iran is calling its drone the "Ambassador of Death".

The Future is Now

Around the midpoint of the first decade of the twenty first century, the year 2005 to be precise, the United States military had fifty drones that could be used at any time. If one were to guess how many drones they had now, one would probably assume the numbers had doubled. A wilder guess would be that the United States military now has ten times the drones that it had ten years ago, which would bring the number up to a whopping five hundred drones. Absolutely no one would assume that the United States military had over a thousand drones now, considering that it had only fifty about a decade ago.

The actual number, however, is far beyond mind boggling. The United States military has a fleet of over 7500 drones that are active and can be used at any time. Whereas the number of unmanned aircraft has gone up by a massive margin, the number of manned aircraft has gone down by about thirty percent. There are over a hundred and fifty drones in the possession of the United States military that can be considered multi-role drones, which means that they are capable of both discreet reconnaissance as well as full blown offensives via missile strikes and bullets. Within the next decade, it is believed that the number of multi-role drones is going to quadruple.

The United States military has a sizeable army of drones. It has a small core of drones that, if they were sent to attack a moderate sized enemy camp, they would decimate the camp, allowing the military to, hypothetically speaking, completely annihilate an enemy base with few, if any, casualties. What does all of this mean? The answer to this question is quite simple.

It is estimated that the worldwide expenditure on drone technology is close to six billion dollars a year. It is also estimated that this amount is going to double within the next decade. Drones are being made to provide soldiers with support, with each soldier possessing a support drone in order to prevent heavy losses. Drones are being made to intercept enemy missiles, to scout ahead and uncover hidden enemy mines and traps, to conduct missile strikes on targets that are further and further away, and to refuel manned aircraft in midair. Drones are becoming cheaper, faster, stronger, more efficient, more autonomous and, most importantly, more lethal. They are becoming the perfect killing machines, they can allow the armies of the world to conduct entire wars without losing a single real soldier and they are here, now, available to buy and research and improve.

What all of this means is that the future is here. Imagining a world in which drones are regularly used in military operations is no longer speculation about the not too distant future, it is a certainty of the present reality that we exist in, it is a facet of the military that is just a step or two away from being the norm.

The Good That Drones Can Do

Drones can be used for so many different things, it's mind boggling. The most obvious use that drones have is, of course, their use in the military, but as you have probably understood from the previous chapter, drones are, more often than not, used by civilians rather than the military.

Drones have a wide variety of uses in the public sector, and the vast majority of these uses are good. Here are some of the good things that drones bring to our lives, and the various ways that using drone technology is improving the way we live:

Drones Help Make Farming Easier

There are few jobs in this world that are as difficult as farming. It requires you to constantly stay on your toes, to be ever vigilant and, most importantly, to be responsible. You are responsible for the food that people are going to eat tomorrow, which is a responsibility that cannot and should not be taken lightly.

Apart from the more philosophical difficulties that farmers face, managing a farm is just, in general, a difficult job! You have to sow the fields, spread fertilizer, pesticides and insecticides in order to ensure that your crops stay healthy, and constantly monitor said crops in order to make sure that they are growing the way that they are supposed to.

Drones can be extremely helpful to farmers in a lot of ways. First and foremost, they can help farmers monitor their crops. Drones are being invented nowadays that have sensors on them, sensors that can monitor, from the air, the substances that the crops are emitting and thus determine whether the crops are healthy or not. Additionally, drones can be used to spread pesticide and insecticide as well as fertilizer from the air, making the whole process a lot easier for farmers.

Drones Can Be Used in Retail

Over the years, people have begun to want things faster and faster. The invention of the computer and the internet has allowed people to attain instant gratification. Things that would have once taken weeks to arrive started to arrive in days, and then in a single day. You could now order food online and get it in less than an hour. An entire industry was built on the delivery of goods to consumers, and much profitability was seen in the improvement of said industry.

Human beings are inherently inefficient, so what better way to improve the delivery system than by incorporating efficient drones into it? Drones can be programmed to travel straight to their destination. They cannot be distracted, they simple cannot get lost, which makes them the epitome of delivery boys.

As has been mentioned before, Amazon has begun to incorporated drones into its delivery system, allowing the company to deliver goods to its consumers within an hour. Fast food companies have also begun speculating about the use of drones in the food delivery industry, as the benefits of using drones in such a way far outweighs the costs.

The Potential of Drone Photography is Limitless

Drone's can be used to reach the most hard to reach places, this much is obvious. You can head to the top of mountains, you can traverse the driest deserts. So much potential can be found just in the way drones can go into such hard to reach places that one can imagine just how useful drones can be in the realm of photography. Drones are so useful in this area that this section of this book will have several subsections pertaining to each area of drone photography as is listed below:

1) The Film Industry: The most obvious situation in which drone photography can be extremely useful is, of course, the film industry. Shooting wide, glorious shots has become extremely popular of late, now that camera technology has improved to such a point and extremely wide lenses are available to enable such amazing shots to be taken. Drone technology takes this to another level. Wildlife can be filmed safely, moving shots of impossible high mountains can be taken, aerial shots of moving actors can be taken, drone photography opens up a whole new realm of possibilities in the film industry.

2) Wildlife photography: Photographing wildlife is an almost inconceivably dangerous job, or at least it can be if the proper precautions are not taken. The fact remains, photographing wildlife is dangerous, not just for the humans doing the photographing the animals but for the animals themselves as well. Wildlife photography can be intrusive to the animals, an invasion of their privacy, and a disturbance that can upset them and result in them being unwilling to go their regular haunts which would result in major upsets in the way they live. Using drones allows photographers to capture their metaphorical prey from a safe distance, whilst not upsetting the delicate balance of the habitats of these animals.

3) Real estate: This can be considered a somewhat unusual entry in this list, because when one thinks of drone technology one usually does not think that the real estate industry would benefit from it. However, if you think about it, photography is very important in

the world of real estate. Think of just how much it would help the real estate industry if people could take a digital tour of a home that they are interested in through the use of a drone!

4) Sports coverage: Sports has become one of the biggest industries in the world as of late. The international football association has become one the richest entities in the world, and basketball players earn obscene amounts of money. This means that people are looking for more and better ways to help audiences immerse themselves in the game. Using drones to improve sports coverage is a great idea. It will allow audience members to get closer looks at the game that is going on, but at the same time it might distract the players from their jobs which is something that must be taken into consideration before incorporating drone technology fully into sports coverage.

5) Police cameras: Keeping the peace is a difficult job, but it becomes especially difficult when you don't know what's out there or what's coming. Often, police officers have to risk their lives attempting to photograph illicit activities taking place. Making use of drone photography can allow police officers to collect the intelligence that they need to put criminals behind bars without them having to risk their lives, something that would put a lot of people at ease. However, at the same time gathering intelligence using a drone may allow the police to infringe on people's civil liberties.

6) Journalism: Photography is one of the most important elements of journalism. Without a visual reference, the news would be just words on paper, or words on your television screen. Photographs and videos help the people watching or reading the news, but are often quite difficult to obtain because the area being reported on is dangerous or inaccessible. Drones allow journalists to bypass these problems and get breathtaking visuals and imagery that will help the people accessing the news source to really relate to the story feel involved in it and moved by it.

Drones Save Lives

Take the drone out of the military and its dubious record becomes suddenly irrelevant. In the public sector, when being used solely by civilians, drones have a tendency to save lives rather than end or endanger them.

Drones have often been used to help distribute food and supplies to victims of natural disasters. Areas ravaged by flood are often inaccessible until the waters have subsided. The same goes for areas that have been affected by hurricanes, earthquakes or any other natural disasters. What all of this means is that the people that have become the victims of these natural disasters often have absolutely no recourse. Help cannot reach them for a long time, and by the time it does reach them, it is more often than not far too late for anything to be done about it.

Drones help greatly in situations like this. Drones filled with supplies can be sent across ravaged landscapes to provide the people that have been affected by natural disasters with food, medicine and water, in short all that they are going to need in order to get through the situation that they are in.

Additionally, there are a lot of countries in which medicine is in short supply. Several African nations, where AIDs and HIV are prevalent, are short on medicine, and any medicine that is sent to these people is, more often than not, stolen by the criminal element that is so pervasive in these regions. This criminal element also makes it extremely dangerous for people to physically transport these medicines as well. Often, an armed escort is required in order to protect these people, and more often than not the armed escort proves inadequate and the medicines are stolen anyway.

Drones can be used to discretely send medicines to these areas, where the people that need them can hide them before the criminal element can get to them. Food and resources can also be provided to these areas, particularly in instances where severe famine or drought has hit.

All in all, drones can be used to help a lot of people. In fact, when you take the drone out of the sphere of military use, it is used to save people almost exclusively. It facilitates the transportation of valuable to resources to the people that need them, and it does so without endangering further life.

Another amazing way in which drones can be used to save lives has been seen recently in Australia, where a drone was used to help defibrillate a heart attack victim. The concept of drone based medical care is still quite new, but the possibilities are exciting to say the least. Drone based emergency medical services can help people get emergency medical care a lot faster than ambulances can. Certain parties have speculated that they can at least arrive before the ambulance and ascertain the situation with the patient so that when the paramedics arrive they are prepared for what they will have to do.

Drones Might Be the Future of Transportation

Drones are already being used very commonly to transport small objects short distances. Theoretically speaking, they can also be used to transport larger objects longer distances as well. The general principle will remain the same. The general operating system and programming will also be similar to what we see in smaller drones. Only the body of the drone will have to be larger and stronger in order to support the objects that it is carrying without breaking down, and the fuel source of the drone will also have to be substantially more powerful.

However, a German company recently successfully made a prototype of a drone that could ostensibly carry people. The sixteen propeller prototype's test drive was successful. It was able to fly for around twenty minutes without crashing. The only real problem with the machine was that its battery is not powerful enough to support long distance flights, and carrying a single person cut its already meager battery life in half. However, it should be noted that this drone is still a very early prototype, and will probably improve drastically in the coming years.

The potential of drone technology in the realm of transportation is, like all other things in the world of drones, virtually limitless. Certain officials have proposed an entire network of public transportation that is remotely controlled. This would allow this whole network to be monitored and controlled by a smaller group of people, as each vehicle would not need its own driver. This has the potential to dramatically decrease the chance of human error occurring during the day to day operations of these vehicles. These hypothetical public transport systems would also be a lot more efficient, as the central control unit of all of these drones would be hard at work organizing the routes. Since each vehicle would not have its own driver, the chance of a delay would also be dramatically reduced.

This would also allow public transport to operate on a twenty four hour basis. In most countries, public transport greatly slows down during the later hours of the night. This often causes great inconvenience for people who work at night or who simply need to get home late at night and cannot afford a taxi. Using drone based public transport can end this inconvenience and provide a consistent method of transport that is efficient, reliable and a lot safer than public transport that is manually operated.

This also ties in with the previous section regarding how drones can save lives. Often, certain situations where people are stuck are too dangerous for a pilot. People might not be willing to risk their lives, or the probability of failure is probably just too high for anyone to be willing to take the risk. In such situations, transporter drones would be infinitely useful. They could be remotely piloted and sent to these areas afflicted by natural disasters and rescue the people that are there.

Drones Can Help Save the Environment

Not enough people care about the environment. This is a hard truth that people need to start understanding. This means that organizations that are attempting to rectify or at the very least analyze environmental damage have very little manpower to work with. Additionally, some situations in environmental damage must be analyzed simply cannot be tackled effectively because it is simply not humanly possible to tackle them in a manner which is efficient. All in all, saving the planet is a difficult job because it requires an extraordinary set of skills to accomplish this enormous task.

Luckily for the people that are attempting to save the planet, drones can help them greatly in their noble endeavors. First and foremost, drones can be used to help reduced pollution. Governments have begun using drone technology to release chemicals into the air that will break down smog and other pollutant gases and turn them into oxygen. The possibilities of such a technology are amazing to behold, and are only possible because of the fact that drone technology existed in the first place, because piloted aircrafts would simply take too much time to distribute such chemicals into the atmosphere.

Apart from such invasive and somewhat hypothetical operations, drone technology is also frequently used to monitor the atmosphere and to detect how much pollution is being released into it. This is extremely important, because in order to reverse the damage that we have already done to this planet we first need to understand the damage that we have already done to it. Drones that possess sensors and the like can be used to monitor the quality of the air that we are breathing and ascertain what measures must be taken in order to make the air more breathable and thus prevent diseases caused by breathing dirty air.

One very effective use of drones in the field of environmental management is being currently utilized by NASA. NASA uses drones to analyze the ozone layer and to ascertain how much damage has already been caused to it due to pollution. The ozone layer is what protects us from cosmic radiation and is an important part of what makes the Earth habitable for us. The use of drones in this manner can help us to prevent further damage to this precious layer of ozone and to prevent the mass climate change that we have been seeing over the past few decades.

One of the more straightforward ways in which drones are used to combat damage to the environment can be seen in Italy, where the government uses drones to monitor sites where illegal dumping of waste occurs. Using these drones, environmental protection agencies are able to gain irrefutable proof that people have been dumping their waste illegally in that area, and are hence able to prosecute these people that were polluting the environment. Overall, the use of drones to save the environment has resulted in a much easier process for the people that work hard to keep the planet a habitable place.

Drones Are Entertaining

Out of all of the possible uses for drones, none is perhaps as limitless as the drone's ability to entertain us. An entire new industry has developed that is based just on the fact that people *enjoy* piloting drones. Within the past decade, drone use for entertainment purposes has exploded, particularly since drone technology has advanced to the point where flying a drone is a lot less complicated and a lot more fun than it used to be, meaning that it has become both more accessible as well as more effective at its job.

Photography via drones has also become a popular hobby for many people, particularly since the use of drones facilitates a far wider spectrum of possibilities for the amateur photograph than a regular camera, no matter many gadgets he adds to that camera. As of late, cameras in drones have become similar to cameras in cellphones: the advancements are coming along quickly, and the new products are marvelous to behold. Additionally, drones have made it easier to make videos for leisure as well. Going on vacation, you are able to record yourself and your experiences with drones, albeit very expensive drones, so that you don't have to worry about taking photographs yourself and can enjoy the experience of your vacation instead.

Entertainment via drones is not restricted to photography, however. Indeed, a lot of people just like to sit in the park and use drones to explore the area around them. Apart from this, just the act of flying a drone has become an increasingly popular activity, much like the hobbyists that built and flew modern airplanes. Drones are becoming an increasingly formidable presence in the world of entertainment, and for good reason considering just how much fun they actually are!

The Other Side of the Coin: The Scary Aspects of Drone Use

No matter how much you are told about the benefits of drone use, the fact remains that you are always going to associate technology with its most sinister uses, mostly due to the fact that these uses are the ones that we see and hear about every day. It is important to understand that, for all their benefits, drones can be used in a dangerous manner, and so it is very important that we understand this dangerous manner in which drones can be used. Hence, in order to facilitate a better understanding of the more unseemly aspect of drones, what follows is a list of reasons why the use of drones should not be encouraged.

Drones Are Not Regulated By International Law

The first half of the twentieth century was a pivotal era for humanity. We had modernized and industrialized, yet the way we negotiated, the way we settled problems with one another was still inherently flawed. This can be seen with two obvious examples, which were the two World Wars that were fought in the first half of the twentieth century. The losses incurred from these world wars were utterly devastating, so devastating that the way our nations interact today is still affected by the aftermath of these wars.

These World Wars are particularly important because the second of these two wars contained the first instance in which in which a modern weapon of mass destruction, a nuclear bomb to be precise, was used when the United States bombed the Japanese cities of Hiroshima and Nagasaki in retaliation for the Japanese attack on Pearl Harbor. The aftermath of the use of this nuclear bomb can still be felt today. Such was the impact of these wars that several international laws were put in place whose sole purpose was to prevent such meaningless loss of life.

In the twenty first century, drone warfare has become the next big thing in the world of the military, and the scary thing is that drones are not regulated by these international laws. Armies using drones take advantage of loopholes in international laws that define warzones and the use of weapons in said warzones and, as a result, are often thought to infringe on the human rights of the civilians that are killed as a result of these drone strikes with impunity. Hence, it is important to look at drones and realize that they are one of the most powerful weapons that are still not fully regulated by international laws put in place to prevent unnecessary loss of life in war zones.

Drones Are Veiled in Secrecy

In any war, losses are incurred. That is just the way that wars work. However, whichever entity is incurring these losses, whether it is the army of a certain nation, a mercenary group or a guerilla group of rebel soldiers, this entity will be held accountable for the lives that they have taken. Most importantly, this accountability means that every life that they do take will be questioned. They will have to justify the killing of each and every person that they kill.

This is an extremely important aspect of modern warfare. A couple of centuries ago, wars were fought with abandon. People were killed, soldiers often sexually violated the women of the country that was in opposition to their own, and in general there was very little concern for human life or for human rights. Modern warfare is a much safer affair, if it is possible for wars to be safe, because it minimizes the slaughter of civilians. The absence of such laws would mean that armies could kill with abandon, something which could often lead to genocide.

Through the use of drones, however, the US military in particular is starting to evade accountability. There is no exact estimate of how many civilian losses are incurred due to drone strikes, and the actual targets of said drone strikes are never specified. It is often speculated that governments have entire kill lists of people that they believe are threats to national security which they claim must be killed, and so justify the use of drone warfare in such a manner. However, the fact that nobody really knows all that much about drones and what they are being used for goes to show that we really need to reevaluate the way drones and especially drone warfare are treated.

Drones Are Not As Effective As You Think

This point ties in somewhat with the previous point regarding the fact that drones allow militaries from around the world to attack their targets without having to account for the actions that they are committing.

When you think of drone warfare, you don't just think of drones bombing enemy bases and camps. There is another thought that probably crosses your mind when you think of such things, a thought that you probably have never questioned before. This thought regarding the use of drones, this assumption as it were, is that using drones is effective. Absolutely no one questions this. After all, if the government is doing this it probably has something in mind right? The militaries of the world would not be using a military tactic that is not effective after all. However, when you look at the statistics you will find out something that is rather shocking about drone warfare, and this is that it is not quite as effective as its widespread use would indicate.

The fact of the matter is, drones don't kill as many radicals and terrorists as it creates. In northern Pakistan, an area that is frequently targeted by drone strikes, terrorists have started using these drone strikes as excuses for their terrorist activities. This goes to show that drone warfare is actually increasing terrorism by exacerbating the sentiment that turns these people into terrorists in the first place. Hence, with drones not as effective as they are widely believed to be, perhaps it is wise to question their use in modern warfare.

The Countries Targeted Using Drone Warfare Are Negatively Impacted

Terrorism is not really the fault of the country in which the terrorists are located. For example, the vast majority of drone strikes that occur happen in North Pakistan due to the fact that a once significant terrorist group called the Taliban are present in that area of the country. However, these terrorists are not actually Pakistani in origin. They originated in Afghanistan, and were essentially former soldiers and insurgents that fought against the Soviet Union and came into power after being armed by the United States. They escaped to Pakistan, mostly as illegal immigrants, when the United States military entered Afghanistan to annihilate the Taliban presence in the country.

In Pakistan, their terrorist activities began to affect the Pakistani public. The rise of terrorism within Pakistan coincided with the arrival of the Taliban within Pakistan. The tens of thousands of people that have died as a result of terrorist activities initiated by the Taliban were killed because of the fact that a terrorist group from a neighboring country entered their country illegally and began to set up shop there, something that the Pakistani government itself acknowledges and is working on eradicating.

However, drone strikes are getting in the way of this. First and foremost, drone strikes exacerbate the terrorist situation that is already present. They fuel the radical agenda and give these terrorists a reason to go about the terrorist activities that they were already so hell bent on. Additionally, these drone strikes damage Pakistani infrastructure. The US military does take the initiative to kill these terrorists but it does nothing to fix the collateral damage that it causes. The cost of rebuilding what is destroyed falls on the government of that nation. In this way, drone warfare is actually negatively impacting countries that have nothing to do with the terrorists that they have hiding on their soil.

Militaries Use Drones for Preemptive Kills

A preemptive kill is an important topic of discussion for anybody that is interested in the military and the way that modern warfare works. This is because the notion of preemptive kills is one that flirts with the very edge of what is moral in an already obscenely amoral state of affairs in which the loss of human life is not only inevitable, it is accepted. A preemptive kill defines the very barrier between what makes wars and military operations the right thing to do or the wrong thing, because these actions that result in the loss of life must be justified.

In order to understand why preemptive kills are so wrong,

one must first understand what exactly a preemptive kill is. A preemptive kill is essentially any sort of military offensive in which a target that is essentially a potential enemy is killed before they have the chance to commit a crime. The key word here is potential enemy, because the target of the drone strike in this situation has not committed any crimes yet.

This goes against everything that is just about fighting terrorism. The law exists for a reason, and it states that no one can be killed with such impunity. The fact that preemptive strikes occur ties in to the fact that military entities that initiate drone strikes have very little accountability. If a target is discovered who has a history of associating with known terrorists or who has been noted to act suspiciously, this person can be targeted as long as the people targeting him are held accountable and have to answer for this action. Preemptive strikes with zero accountability can spell an end to warfare with minimal losses, and is thus a scary possibility to ponder.

Drones Make it Impossible to Undertake Peaceful Dialogue

There is a general consensus among humanity regarding the state of the human psyche in today's day and age. In general, we believe that we are more educated, more civilized and just all round better than we were perhaps a century or so ago. The most important part of this more civilized society that we live in is, perhaps, that our attitudes towards warfare has greatly changed. No longer are we the kind of people that invade other countries with impunity, nor are we the kind of people that, when wars occur, take human life with abandon and completely annihilate our opponents with no thoughts towards who they were or where they are coming from.

In general, we pride ourselves on our ability to look at each other as human rather than as objects towards which we can direct our hate or our vitriol. We are, in general, a more acceptant society, and this belief, unlike so many other beliefs that have been disputed in this book, is completely true. We *are* a more acceptant society, we have learned to humanize one another.

This is important to note because drones are extremely dangerous to this notion of peace that we have come to associate with our modern society. The peaceful dialogue that is so important to the way we resolve issues is slowly being abandoned in favor of killing the ones whom we consider our enemies from afar. This issue will not seem all that important to the ones doing the attacking, but if we are truly the civilized people that we believe we are we will attempt to humanize the other and favor dialogue and discussion over remote controlled bombing.

Drones Kill Civilians

This is perhaps the single most important point that one can make in this situation. After all, what is more important than human life? Our entire existence, our entire purpose for being is to safeguard the sanctity of human life. After all, this is what makes us human. This is what separates us from the animals and from the people that we once were. This is what justifies the fact that we like to call ourselves more civilized, more educated and just more human than we were not all that long ago.

There is absolutely no doubt that drones kill civilians. An entity that is invested in drone warfare, such as a company that creates drones or a military organization that benefits from the use of drone warfare, would say that collateral damage is a part of any military operation, and that it should be accepted that we are going to lose people in our quest to stay the hand of terrorists that want to kill the people that we love. War is, after all, a numbers game. As long as the collateral damage does not exceed the number of lives that are saved, the commanders of the military organization will consider the military operation a complete success.

However, the fact that drones kill civilians ties into the fact that the people that use drones are, for all intents and purposes, accountable to no one. They do not have to answer for the civilians that they have killed and they are often vague if they are forthcoming at all about how many civilians are actually killed. This means that there is no accurate number that can tell us how many civilians are actually killed during these military operations. If any military entity in the world is allowed, in any way, to kill civilians and not have to answer for it, there is something really wrong with the situation, in this case the use of drone technology, which enables them this level of freedom.

Drone Warfare Has Very Little Controlling It

There is an entity within the United States that oversees the program. This entity is called the United States Congress Intelligence Committee, and technically it has the power to make the people that are a part of this drone program, the people that actually decide who dies and who lives and what is an acceptable level of collateral damage, answer for what they are doing and justify their actions.

However, there is a slight, and that word is used ironically, problem with this committees control over the drone program: the program is classified. This means that everything that occurs as a result of this program is a complete secret, even to the entity that has been given the authority to control and oversee this program in the first place. There is something terribly wrong with any military program that has the ability to take lives in large quantities and exercises this ability frequently if its entire functioning and inner works is hidden from the entity that is supposed to be regulating it.

The reason this is so dangerous is simply because the drone program has the potential to wipe out small towns if it is used effectively and in concentration. If the Congress Intelligence Committee is not told about what the drone program is up to, it can claim plausible deniability. They can simply claim that they did not know what was going on due to the fact that the program was confidential and classified.

This must be rectified if drone warfare is to come out of the darkness and begin to enter the realm of both legality as well as morality.

Drones Have Become Part of a Troubling Narrative

The United States of America widely and famously touts itself as the greatest nation in the world. This mentality began after the victory of the Allied nations in the Second World War. The United States had a superior military to nearly every other country in the world, and it was also quickly becoming the land of opportunity. Immigrants from around the world were heading to the United States in order to find their fortune, and the fact of the matter is that a huge amount of them actually did.

America also proudly calls itself a protector of civil liberties. This was actually the foundation of what America was built on, the core values of the United States as a single nation. However, in recent times America has begun to be viewed with a lot more trepidation by other countries in the world. America's foreign policy is seen as a modern variation on imperialism, with America attacking countries with abandon as though it has nobody to answer to, and the fact of the matter is that this is true to some extent.

America's military policies have also begun to be heavily criticized. The use of torture techniques and other violations of human rights on terrorists and enemy soldiers has been criticized as being unbecoming of a nation that calls itself the greatest in the world, and that regularly promotes itself as the only nation in the world that protects civil liberties. This shift in narrative means that America has started to be seen as more of a bully than a protector, as something that is to be feared rather than loved and respected.

This is actually a very sad state of affairs. America was once the greatest country in the world, and very few people would dispute that. Now it is one of the most feared countries in the world. Drones have become a huge part of this shift in narrative. No country in the world uses drones as much as America does. This killing with abandon, without caring for the consequences of their actions, this is what is turning the world against America. All the US military needs to do in order to reverse this shift in narrative is to be more forthcoming about at least the drone program, so that they at least look as though they care about human life.

Drones Can Be Used to Invade Privacy

This is, perhaps, the only point in this entire list of the negative aspects of drone use that has at least something to do with civilian use of drones. However, the non offensive use of drones also plays a huge part in the fact that drones are making it a lot easier for entities, whether the entity is an independent person or a military or intelligence gathering agency, to spy on people and to invade their privacy.

In the realm of civilian life, drones can be used to spy on people just to cause mischief. One of the most important situations in which this happens is the realm of celebrity photography. People that photograph celebrities are usually called paparazzi, and are most often considered a nuisance. This is due to the fact that celebrities often feel harassed by these photographers. Often, celebrities are photographed while simply going out to eat in comfortable clothing, and end up seeing a scandalous headline attached to this picture the very next day.

Drones are making it a lot easier for these paparazzi to do their job, and that is not a good thing at all. Using drones, paparazzi can even enter the homes of certain celebrities in order to get pictures that would be as scandalous as possible. One can only imagine how much of an impact this can have on the mental health of the people that cannot get a moments peace in the first place.

Additionally, the enormous increase in drone use is also worrying because a third of all drones being acquired are being acquired by the Central Intelligence Agency, and they are being acquired for the purpose of being used on American soil. The use of drones will enable intelligence agencies to spy on the citizens of their own country and to gather information. In short, privacy will become nonexistent if drones are used in such a manner more commonly in the future.

Drones: Good or Bad?

This is an important question to ask. After all, drones are not going anywhere. They are where they are, and instead of becoming less prolific they are most probably going to become more and more commonly used over the next decade, the point where the US government will probably have tens of thousands of drones at its disposal.

Additionally, it is important to note that drone use in personal spheres is also becoming increasingly common. Within the next decade, there might be hundreds of thousands of drones that would be used for personal purposes, whether for photography, just as a hobby or for more unseemly purposes.

What all of this means is that drones are here to stay, and are only going to become an ever more common presence in our day to day lives. Hence, in order to understand what to make of all of this we need to ask ourselves, are drones good or bad?

Before this question is answered, you must first notice one very important aspect of all of the information that you have been given above. This single fact will tell you a lot about this question, and will probably help you come up with an answer yourself.

This fact is that the vast majority of the negative aspects of drone use had to do with the way they were being used by military entities around the world, essentially all violent uses of the drone that involved the taking of human life or the taking of civil liberties.

Conversely, the vast majority of the positive uses for drones had to do with the way they were used by civilians and corporate entities within the private sector. Most of these uses had to do with helping make people's lives better rather than taking life away from them.

This is not to say that all military use of the drone is bad, or that all civilian use of drones is good. After all, using the drone allows the military to prevent the loss of human life as long as drones are not used violently. Additionally, a paparazzo photographing a celebrity and invading his or her privacy comes under the use of drones in the private sector, and it is pretty obvious that this is an unseemly use of drone technology and is certainly nothing that should be encouraged or promoted in any way, shape or form.

However, the fact remains that when drones are used in a non violent way, in a way that does not invade anyone's privacy, they are wonderful machines. They can be used to help people, and are a firm and solid part of the recent new wave of technology.

This new wave of technology refers to all technology released in the twenty first century, all electronic media essentially, that is essentially connecting us all and is helping us to help one another. Drones are just as important as smartphones in helping make the world better connected, and in helping us reach people that were previously unreachable.

The pros and cons regarding the use of drones is before you. Drones save the lives of friendly soldiers, but take the lives of civilians. Is this a worthy exchange? Is the taking of human life with impunity *ever* okay? Can it ever be justified this way?

Perhaps if the military use of drones was better regulated, perhaps if it did involve at least some level of accountability among the people that were launching these drone strikes and taking human life, the use of drones in a military setting would be justified. On the other hand, perhaps war conducted through machines, war in which you are not part of the dirt and blood and sweat that is the battlefield is not a good idea for anybody. Perhaps it would make us irrational, more prone to violence. However, these questions are purely philosophical in nature.

Drone technology cannot be bad. It does too much good to be bad, it saves too many people in need, makes life too easy for it to be bad. However, the fact remains that until the most famous use of drone technology is not completely eradicated, the use of drones to take human life, drones can never be considered completely good either.

Conclusion

While drones aren't new, they continue to be modified with updated technology. Additional legislation is being incorporated to try to keep up with the advanced technology, but continues to lag behind.

Drones are here to stay. Individuals have the right to speak up about their concerns. However, citizens also have a responsibility to be educated and well informed on the topic.

The FAA and other organizations, including local governments, are working hard to come up with laws that will hold people accountable with drone use. Serious fines and even jail time for violating laws and privacy are being considered.

I hope this book has helped educate you and bring you a greater understanding of drones and their uses.

RECOMMENDED READING

MINIMALISTIC LIVING: How To Live In A Van and Get Off The Grid

hyperurl.co/offthegrid

Boundaries: Line Between Right And Wrong

hyperurl.co/boundaries

SELF ESTEEM: Confidence Building: Overcome Fear, Stress and Anxiety: Self Help Guide

hyperurl.co/selfesteem

ROME : Ancient Rome: Roman History and The Roman Empire

hyperurl.co/rome

Made in the USA
Middletown, DE
09 March 2016

29989219R00055